WIC 20th Anniversary: Celebrating Women in Chemical Engineering

Topical Conference at the 2018 AIChE Annual Meeting

Pittsburgh, Pennsylvania, USA
28 October - 2 November 2018

ISBN: 978-1-5108-7633-0

Printed from e-media with permission by:

Curran Associates, Inc.
57 Morehouse Lane
Red Hook, NY 12571

Printed by Curran Associates, Inc. (2019)

For permission requests, please contact AIChE
at the address below.

AIChE
120 Wall Street, FL 23
New York, NY 10005-4020

Phone: (800) 242-4363
Fax: (203) 775-5177

www.aiche.org

Additional copies of this publication are available from:

Curran Associates, Inc.
57 Morehouse Lane
Red Hook, NY 12571 USA
Phone: 845-758-0400
Fax: 845-758-2633
Email: curran@proceedings.com
Web: www.proceedings.com

TABLE OF CONTENTS

Some abstracts in this book refer to images that were not made available by the publisher

AIChE Women's Initiative - Our Past, Our Future

by Caroline C. Reynolds, CR Solutions, Austin, Texas

We will begin by looking at women's history in AIChE and the beginnings of the Women's Initiative Committee, then we will turn to the future and think about what we would like to accomplish in the future.

Twenty years is a long time, and we have seen many changes.

This has been an exciting year, because I have been working to elect a woman mayor of Austin. She always says that she is a mathematician who was a project manager at Lockheed. However, many people introduce her as an engineer. It makes me proud that people think that a smart woman who understands budgets must be an engineer.

I doubt that anyone would have thought any woman was an engineer fifty or sixty years ago.

When I started at Rice in 1960, I had no idea about how many chemical engineers were men or women. I'd never met one, but my father had and he assured me that I could be an engineer. My first two years in college were spent taking chemistry courses and I met and dated two guys who were going to be chemical engineers.

At that time Rice was the press center for NASA, and all the NASA astronauts were men, now there is a diverse team of men and women, and each of these three are chemical engineers.

Figure 1. Recent Astronauts – Dr. Mary Ellen Weber, Dr. Mae C. Jemison, and Dr. Robert L. Satchler, Jr.

Dr Mary Ellen Weber

Dr Mae C. Jemison

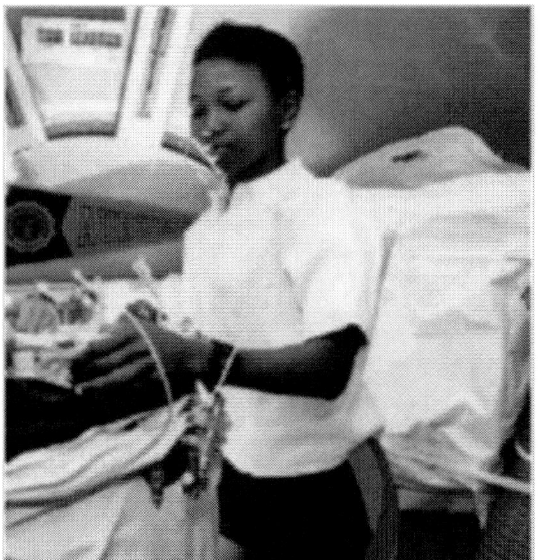

Dr Robert L. Satcher Jr

If I had not changed my major to Chemistry in my junior year, I would have been one of less than 23 women engineers to graduate in the US in 1965. When I completed my Masters in Engineering in 1979, only 15 years later, I was one of only 120 women who did so. That was a 5-fold increase. Surprisingly, more than 1000 women received Bachelor Degrees that year. There was a 10-fold increase in women getting Bachelor degrees in less than 20 years.

Table 1 includes data from NSF and shows the enrollment growth of women in Chemical Engineering between 1966 and 2004.

Table 1. Chemical Engineering Degree Recipients by Level and Gender

Chemical Engineering Degree Recipient by Level & Gender

Academic year ending	Bachelor's Total	Bachelor's Men	Bachelor's Women	Master's Total	Master's Men	Master's Women	Doctorate Total	Doctorate Men	Doctorate Women
1966	2,981	2,958	23	1,072	1,065	7	367	365	2
1967	2,997	2,969	28	1,028	1,021	7	330	327	3
1968	3,395	3,365	30	1,251	1,237	14	377	374	3
1969	3,768	3,715	53	1,227	1,216	11	422	419	3
1970	3,995	3,938	57	1,127	1,109	18	457	454	3
1971	3,907	3,843	64	1,200	1,173	27	407	403	4
1972	3,967	3,887	80	1,259	1,230	29	391	389	2
1973	3,968	3,874	94	1,139	1,117	22	424	420	4
1974	3,826	3,706	120	1,111	1,080	31	418	409	9
1975	3,420	3,273	147	1,078	1,051	27	396	391	5
1976	3,543	3,254	289	1,129	1,088	41	335	327	8
1977	3,986	3,534	452	1,179	1,110	69	329	319	10
1978	5,205	4,453	752	1,335	1,245	90	282	277	5
1979	6,442	5,387	1,055	1,276	1,156	120	315	306	9
1980	7,276	5,989	1,287	1,393	1,249	144	316	302	14
1981	7,639	6,274	1,365	1,406	1,230	176	317	306	11
1982	8,059	6,447	1,612	1,409	1,222	187	333	314	19
1983	8,550	6,761	1,789	1,545	1,369	176	392	369	23
1984	9,192	7,115	2,077	1,798	1,590	208	409	382	27
1985	8,941	6,848	2,093	1,814	1,529	285	504	463	41
1986	7,411	5,805	1,606	1,641	1,401	240	531	470	61
1987	6,114	4,574	1,540	1,386	1,143	243	584	524	60
1988	4,654	3,522	1,132	1,322	1,107	215	685	620	65
1989	4,187	3,017	1,170	1,321	1,092	229	712	632	80
1990	3,834	2,745	1,089	1,205	1,013	192	658	580	78
1991	3,728	2,564	1,164	1,025	852	173	691	605	83
1992	4,123	2,854	1,269	1,145	914	231	725	609	113
1993	4,899	3,335	1,564	1,220	996	224	737	638	94
1994	5,636	3,953	1,683	1,287	1,008	279	725	609	113
1995	6,391	4,367	2,024	1,369	1,063	306	708	597	109
1996	6,708	4,537	2,171	1,416	1,110	306	798	653	143
1997	6,977	4,701	2,276	1,345	1,013	332	767	641	122
1998	6,721	4,525	2,196	1,372	1,028	344	776	630	140
1999	NA	NA	NA	NA	NA	NA	674	550	123
2000	6,219	4,016	2,203	1,352	1,008	344	725	571	151
2001	6,088	4,033	2,055	1,368	1,016	352	729	547	180
2002	5,946	3,823	2,123	1,267	932	335	705	527	176
2003	5,545	3,532	2,013	1,365	1,022	343	648	492	154
2004	5,185	3,352	1,833	1,506	1,089	417	723	549	173

Chemical Engineering Degree Recipient by Level & Gender

In 1966, only 23 women in the US received Bachelor degrees, 7 received their Masters and 2 women were granted PhDs. No wonder I had never seen a woman chemical engineer.

By the late 70s, women have broken out of the single digits in Doctorates, significantly more than 100 had earned their Masters and over 1000 received Bachelor degrees.

My Transport lab instructor was Marge – the first woman chemical engineer that I ever saw. By the mid-90s, there were consistently over 100 women earning PhDs, 300 women received Masters, and over 2000 women were awarded Bachelor degrees. In 2004, the last year for which NSF published figures, there were 1833, nearly 2000, women receiving Bachelor Degrees, 417 Masters Degrees and 173 PhDs !

In the same period, the number of men graduating had grown from nearly 3000 in 1966 to a high of over 7000 in 1984. Since then the number of men graduating has dropped as low as 2500 in 1991, but is now stabilizing in the mid-3000 range. Since the '60s, men have consistently earned about 1000 Masters degrees and 300 to 600 PhDs.

So, women and men earn almost equal numbers of Bachelor degrees, but there is room for improvement in the number of Masters and PhDs. On the other hand, some women (and men) have opted for MBAs or moved to medicine or other areas that don't require advanced degrees.

Table 2 is a look at women in the academic pipeline.

Table 2. Women in the Academic Pipeline

*Table 2. Women in the Academic Pipline**

Discipline	Students				Departments 1-100 FY 2007			
	BS 2004	BS 2005	PhD 86-95	Phd 96-05	ASSIST	ASSOC	PROF	ALL
Chemical Engr	35.6%	36.7%	17.1%	23.7%	24.2%	17.6%	7.3%	12.6%
Civil Engr	24.1%	23.9%	12.7%	22.0%	24.7%	14.5%	7.1%	13.0%
Electrical Engr	14.0%	12.9%	8.6%	12.3%	15.5%	12.5%	5.7%	9.5%
Mechanical Engr	13.7%	13.2%	7.3%	8.4%	18.0%	11.9%	4.4%	8.8%
Chemistry	51.0%	51.7%	26.3%	32.4%	21.2%	19.6%	9.7%	13.7%
Math	46.1%	44.9%	22.5%	28.7%	26.8%	18.4%	7.1%	12.9%
Computer Sci	24.7%	22.0%	19.8%	21.2%	20.0%	11.6%	10.3%	13.2%
Astronomy**	41.5%	42.4%	15.2%	22.7%	25.3%	21.6%	12.3%	15.8%
Physics	21.6%	21.1%	10.8%	14.3%	16.8%	13.4%	6.1%	9.1%
Biological Sci	62.5%	62.2%	39.6%	46.3%	35.0%	30.0%	17.4%	24.4%

*Females were 50.7% of the 2006 US population. **Top 40 departments.

In the early 2000s, women were earning just over 1/3 of Bachelor degrees, and nearly a 1/4 of PhDs. Chemical engineering was topped only by Biology, Chemistry, Mathematics and Astronomy. Biology tops all other STEM areas in the percentage of PhDs, but Chemical Engineering is only slightly less, and similar to Chemistry and Math. Biology still stands out with women in 17% of full Professor positions, with Astronomy and Computer Science in second and third place.

Similarly, in AIChE Membership, shown in Table 3.

Table 3. AIChE Membership by Gender

AIChE Membership by Gender

Year	Female	Male	Unknown	Grand Total
1980	1097	27106	2505	30708
1981	1468	29204	2560	33232
1982	1892	31338	2609	35839
1983	2402	33676	2653	38731
1984	2984	36045	2703	41732
1985	3686	38536	2778	45000
1986	4642	41717	2899	49258
1987	4500	41227	1529	47256
1988	3887	39647	741	44275
1989	3621	38535	329	42485
1990	3583	38202	48	41833
1991	3806	38082	18	41906
1992	3917	37435	15	41367
1993	4616	37521	6	42143

AIChE Membership by Gender

1994	5937	40245	5	46187
1995	7289	43162	5	50456
1996	8504	46169	7	54680
1997	9724	49052	8	58784
1998	9115	47497	14	56626
1999	8849	46867	25	55741
2000	9172	47482	23	56677
2001	8913	46168	25	55106
2002	8599	44265	28	52892
2003	8005	40786	31	48822
2004	6247	33466	18	39731
2005	7503	36446	24	43973
2006	5855	31248	231	37334
2007	7010	33115	1461	41586
2008	6432	31880	1266	39578

In 1980, women made up less than 4%, or fewer than 1200, of the nearly 31,000 members of AIChE. At the same time, I was an officer and Chair of the Austin/San Antonio AICHE section in the early 80s. There were only a few women members, most had been undergraduates at the University when I was a graduate student. By the mid-80s, when I began coming to national meetings, women were about 10% of AIChE membership, echoing the increasing numbers of women graduates and employees. There was a drop in the late 80's downturn, but it quickly rebounded in the early 90s to the low teens and grew to over 9000 women, or nearly 20% of the membership by the late 90s. It was about 15% ten years ago.

In the late 1980s, Maria Burke, Carol Hall, and Alice Gast met for coffee with a few other women to discuss how to help women in chemical engineering realize their potential. They discussed mentoring, networking and room-sharing at AIChE meetings. Room-sharing may surprise you on this list, but it was a real problem, because the US and state governments would not cover the current room rates charged at meeting hotels.

I'd run into a similar problem when, in 1980, I started working for the Texas Department of Health. The state expected men to share a room at conferences and retreats. A few months after I began working there, I was to be the main speaker on the new RCRA Hazardous Waste Rules at an agency retreat, but the Texas allowable rate meant that I'd either have to share a room with one of the men or personally pay for part of my own hotel room. I spoke to my boss, a retired Air Force colonel and told him that it wasn't fair for me to have to cover my own hotel bill to speak to Department employees and I didn't think it was proper for me to share a room with one of the men. He took care of getting the exception so that the state paid for my whole room rate.

The irony in this situation was that, because of my advanced degrees in both Chemistry and Chemical Engineering, I was the only person at the Texas Department of Health that they deemed qualified to review the entire proposed Hazardous Waste (RCRA) rules and draft the Department's comments concerning the classification and testing of wastes.

It was also in the early 80s that I shared a king room with two Austin members at a national SWE meeting in New York City, where the room rate was $200, I believe. By splitting the hotel rate three ways, we had per diem left to pay for our meals!

But back to the AIChE plans…
In the late 80s, Carol Hall organized a reception for women faculty and graduate students and began to keep a list of interested women. NSF picked up the tab for a couple of years and then decided they could not continue, because food was served. DuPont came through for a few more years.

Also, in the late 80s, AIChE named the first two women Fellows, Patsy Chapplear and Elizabeth Drake. Patsy was active in the Houston AIChE section and was an AIChE Director from 1985 to 1987. Elizabeth had taught at MIT and was AIChE Treasurer from 1988 to 1990. When Dr. Drake received and read her Fellow plaque, she commented that it used only male pronouns. It was quickly replaced.

In the early 1990's, Carol Hall and Esin Gulari chaired a few sessions on women's issues and soon after AIChE convened a group to look into addressing these issues in a more organized way.

WIC was established as a committee in 1998.

Meanwhile, I had focused on being Treasurer and eventually Chair of the Environmental Division and my evolving job. I didn't realize that women's issues were being discussed. There was one other woman on the Environmental Division board – a 100% improvement over what I'd known for my whole life, being the only woman engineer in the room. The men on the board were welcoming and encouraging, so I saw nothing amiss.

Over WIC's last 20 years there have been changes – in the size of the Women's Initiative Committee and the things we've been able to do and in how we have funded the committee.

I think it is interesting to look back to see how we started and advanced to this point. Later, we'll ask where to we wish to be in 5 and 10 years from now.

I found the minutes for the November 2002 WIC Executive Board meeting in Indianapolis. It included a list of the officers and the part they played. We did all of our own programming, working from our own experiences and learning. Each of the 15 board members were involved in pulling together WIC activities. We didn't have a template or a routine. I think that you would recognize many of the names today. Some are still involved in WIC today, while others have moved to other endeavors.

However, in 2002, Dr. Christine Schmidt (UTexas, UFlorida) invited me to attend a workshop provided by COACh, then a part of ACS. It was associated with the AIChE national meeting in Indianapolis. When I researched the workshop, I found that it was focused on faculty and thought I didn't fit the focus group, but Christine said that all the places were not filled and that I would probably find the material usable.

The workshop discussed negotiation strategies that women faculty could use to obtain resources, better salary and larger, better-situated research space and offices. This was the first time I'd heard these issues articulated and discussed... and I sat in a room of about 15 women Chemical Engineering faculty and heard their stories.

I believe 2002 was also the first year that Kraft sponsored the WIC luncheon and networking time. They agreed to continue into 2003. And then, for ten years, Kraft continued to give WIC $5000 per year for the luncheon, until 2012, when they became a part of Heinz. This was a significant support when we had a budget of nearly zero. The President of Kraft also sent guardian angel Mike O'Brien to offer suggestions and see how things were getting along.

I really want to give a shout out here to WiC's early AIChE laison, Betty Feehan. She was a mathematician, with an advanced degree in psychology, and she believed in and championed the ideas of WIC. She was always able to find money in the national meeting budget for a couple of bottles of wine and a fruit and cheese platter for the Monday afternoon Executive Board meeting. She helped us locate resources in the NYC offices.

She developed and delivered talks about the Myers-Briggs test to mixed sessions and how to use it to understand your strengths and how to manage others. (I don't want to leave this point without pointing out that Kathleen Cook Briggs and Isabel Briggs Meyers are mother and daughter followers of Carl Jung.)

In these early years we typically had 25 to 30 attendees at the luncheon, which was free with the Kraft sponsorship. The attendance at our workshops was generally around a dozen.

AIChE national office introduced a check off box for members to donate $10 to Committees like MAC and WIC, and we suddenly began to receive nearly $10,000 per year. This enabled us to reserve a room and pay luncheon speakers' travel expenses. We soon came up with the idea of giving some of that money to women to make their first trip to an AIChE meeting and WIC luncheon.

Ten years later in 2012, we had about 60 or 70 members and friends at the luncheon in Nashville and several dozen at the workshops.

In the intervening ten years, I had managed to find the name of the group that had presented the negotiation workshop in Indianapolis. It was COACh, a group formed by the women of the American Chemical Society.

COACh is based at University of Oregon and works to increase the number and career success of women scientists and engineers. It's popular career building workshops have been attended by over 20,000 scientists and engineers (men and women) over the past 20 years.

COACh workshops are available on the topics of

- effective negotiation,
- communication techniques,
- successful mentoring,
- leadership skills,
- publishing in peer reviewed journals,
- grantsmanship,
- memorable scientific presentations and
- launching a career in science and engineering.

COACh is funded by NSF and invested in their staff of expert presenters and resources and this allowed COACh to offer free or nearly free presentations at AIChE meetings many times in the past. I renewed AIChE's link to COACh and they have been a strong supporter of WIC's career building workshops.

Now, I'd like to shift gears and think about what WIC would like to be in the future. WIC started as a gathering place for women chemical engineers, and has become a source of career building sessions and workshops and a source of career-starting travel scholarships. We have increased the number of women applying to become AIChE Fellows and we have taken a hard look at the issues of women's retention and return to work.

Where do we want to fly to in the near future?

My idea is that we should increase the footprint and visibility of women chemical engineers.

How do we do that? I believe we become more visible as we participate in our communities.

I started small by becoming treasurer and repair chairman of the co-operative nursery school, because I was the one used to handling money and tools!

Join your AIChE section and volunteer to be an officer. I was soon the Vice Chair and Chair of the section. This is a great career builder, with opportunities in leadership and planning. Everyone wants to see you succeed and will share lessons learned.

As Chair of the AIChE section, I volunteered to became a judge at the local science fair and pulled the whole Austin section with me into the enterprise. I talked them into providing a prize for the project that best exemplified chemical engineering.

The first year, it was a young man who calculated the reaction constant for the destruction of Vitamin C in spinach in a pressure cooker! It looked like a cooking show, but was actually a complex problem for an independent student project. At the same time, I got the Association for Computing Machinery to join us in this endeavor. That wasn't too hard, because I was married to a long time member.

Volunteer at local schools and be a PTA officer, if your child is a student. I did a magnet demonstration for the second grade class. I organized fund-raisers and obtained donations of refreshments for the Olympics Day. Don't let the description fool you – these are all builders of leadership and project management skills. They also show the community what chemical engineers do well.

A middle school teacher asked me to teach a five-day course on "the environment". I focused on teaching students to read and summarize environmentally focused articles and to write questions for a panel of three environmental guests from government and consulting. I wrapped up with two EPA films on work place safety and chemical carcinogenesis. For homework, I asked the students write questions that they might expect the movies to answer. My presentations gave the teacher a break and provided my an opportunity for more management, leadership and planning. Also, this provided more visibility for engineering to students and the teachers.

I fleshed out and organized a SWE Engineering Workshop for the local Girl Scout council. It was a day-long adventure, ending with the University of Texas Physics Circus. There were 100 Girl Scouts, 25 Scout leaders and 40 engineers. My reward was hearing a leader say, "I just realized that I could have been an engineer." I replied, "You still could be. Go to the University of Texas."

With the Girl Scout Council, I developed requirements for a patch that girls could earn by teaching a skill learned at the workshop to another troop. I designed the patch, using a new bridge over the Colorado River that required the work of Civil, Chemical, Mechanical and other engineers.

You can work within AIChE to judge the ChemE Car Contest and cheered on the participants in the University car building contests.

Work at University on the student retention program and the summer engineering camps.

Work in a school to encourage and support students entering the local science fair. It only requires simple things like providing articles and books and encouragement to the students. Don't expect miracles, but expect to see the number of students to slowly increase and some will win awards.

The best part is seeing them proudly standing with their project and explaining it to a judge. This has taught me more patience and teaching skills.

Work with a teacher to start a Science and Engineering Club at the school. It could be place where you could show short videos on interesting subjects like why and how arches work and show the students how to measure pH with red cabbage.

If you have the expertise, review and comment on proposed changes to City Code and keep the City staff on the straight and narrow. I am often involved to explain how Critical Environmental Features may endanger the structural integrity of homes built nearby. I review staff's use of models to obtain the "right" answers for builders and developers and then develop a short, but effective, presentation to the members of the board or commission who will need to deny or revise the applicants plans. In Austin, with recent increases in impervious cover and higher 25- and 100-year flood elevations, this has become a matter of life and death.

Form relationships with City Council Members who may not have the expertise to understand flooding and traffic issues. Austin is having traffic problems caused by improper signal synchronization and a shortage of bridges across the river at rush hour. Traffic is like pumping an incompressible fluid and I recall Dr. McKetta's Rule of Thumb that you cannot economically pump a fluid faster than 10 feet per second.

As I finish, I want to reiterate that no community activity is too small or unimportant. Each time you identify your self as a chemical engineer and participate in your community, you are a model for a child or an adult to know what an engineer looks like and what an engineer can do.

Start small when you have only small bits of time. Each community activity, whether in schools, or neighborhood association, or professional society, will increase your career skills in planning, management, procurement, and leadership. There is no failure. It is a win for you and for your community.

I hope each of you will begin by trying your wings on small projects and be sure to identify yourself as a chemical engineer.

541547 A Unique Chemical Engineering Career in the Energy Industry

Abstract:

Energy production and demand has changed drastically in the last 25+ years and so has the role of women in the business of energy. Learn how one female, Hispanic Chemical Engineer has navigated a diverse career of unique positions in this ever-changing and high-profile business.

542374 From Supercritical Fluids to Ionic Liquids

Abstract:

One of the keys to efficient and selective separation processes is the ability to tune the power of the solvent and its interactions with the species being recovered. This places the emphasis on the design of materials used to facilitate the separation. Materials design for separations provides an alternative to energy-driven separations like distillation. Supercritical fluids and ionic liquids are two classes of materials that afford this tunability. Supercritical fluids, like carbon dioxide, can be tuned by varying temperature and pressure, which can exquisitely tune the density, and correspondingly, the solvent power of the fluid. Supercritical fluids can also be tuned by additions of small amounts of cosolvent that â€˜clusterâ€™ around dissolved solutes. We will show how both of these techniques have been used to facilitate separations using supercritical fluids. Ionic liquids can be tuned by careful choice of cation, anion and substituents. They allow the design of solvents with multiple functionalities, ranging from polar to nonpolar, that can be extended into reversible reactions. We will give examples of the design of ionic liquids for a variety of separation processes.

541943 How to Grow Your Career While Balancing on One Foot

Abstract:

Women often feel they are not qualified for a role or a project until they meet all the criteria listed. My talk will discuss some of these common female misconceptions, my personal challenges and experiences, and how sometimes the best path to grow your career is somewhat of a balancing act.

541814 Stratification in Colloidal Films and Lessons from Soft Materials for Women in STEM

Abstract:

Multicomponent films based on colloidal dispersions have a wide range of applications, including antimicrobial coatings for medical instruments, conductive textiles for flexible electronics, anti-reflective coatings for optical devices, paints for humid environments that are resistant to mold growth, and drug-loaded coatings for medical implants. Often, there is a need to spatially control location of certain components in the film. In this talk, I will present our recent results from atomic force microscopy (AFM) and small-angle X-ray scattering (SAXS) on films prepared from binary colloidal dispersions containing large and small particles of varying size and initial volume fraction. Our results show evidence of different types of stratification behavior, including large-on-top (e.g., large particles migrating to the top surface of the film), small-on-top, and â€œsandwichâ€-like layering. Finally, in celebration of WICâ€™s 20thanniversary, I will conclude the talk with some lessons I have learned from watching soft materials respond to stress and strain for the past 20 years, which I believe can also be applied to women in STEM who may be under stress and strain. Namely, I will briefly describe the importance of building strong networks and how letting things slide can sometimes lead to greater strength.

541747 Vapor Deposited Polymers: From Fundamentals to Commercialization

Abstract:

Chemical vapor deposition (CVD), as practiced by the semiconductor industry, typically utilizes high powers and high temperatures to drive non-selective chemistry. These aggressive conditions are incompatible with reactants possessing fragile organic functional groups. However, employing selective chemistry allows deposition rates of CVD organic films to be high, even when energy input is low. The CVD method is ideally suited for insoluble and infusible materials such as fluoropolymers, crosslinked organic networks, and conjugated semiconducting and conducting polymers. To date, a portfolio of >70 CVD homopolymers and copolymers have been demonstrated. The conformal nature of CVD polymerization enables the facile integration of organic thin films into device prototypes onto thermally sensitive and mechanical flexible substrates. Scale up of the process has facilitated the commercialization of CVD polymer technology.

542376 Assembly Engineering of Complex Colloidal Crystals

Abstract:

Advances in computer software and hardware now permit real-time discovery, design and engineering of complex materials. In this talk, we show how GPU-enabled molecular dynamics and Monte Carlo simulations, extended thermodynamic ensembles that support "digital alchemy" for rapid optimization in design space, workflow and data management software, fast analysis codes that include autonomous local environment detection, and machine learning algorithms have advanced the discovery, design and engineering of colloidal crystals, providing blueprints to guide the synthesis and fabrication of structures of extraordinary complexity and diversity.

541940 Some Historical Information and Statistics on Women in Chemical Engineering and in AIChE

Abstract:

Women have been in the field of chemical engineering for many decades, but their numbers were quite small until the 1970â€™s. Their role in industry, academia and in AIChE will be the topic, with a look at trends and a discussion of how perceptions have changed with the times, including highlights of some of the leaders in the field today.

541530 Modeling and Simulation of Complex Particle-Laden Flows

Abstract:

Particulate processes pervade the petroleum, chemical, mining, pharmaceutical, and agricultural industries. Many of these processes have significant opportunities for optimization and productivity enhancements. Reliable particle modeling and simulation tools can improve critical understanding and design of particle transport processes; this talk will provide several examples in this regard using both discrete element method simulations and computational fluid dynamics. Specifically, this presentation will focus on complex particulate flows such as those involving non-spherical particles, particle breakage, particle flexibility and cohesion.

541546 Computational Design of Peptides to Detect Human Health Biomarkers

Abstract:

The goal of this research is to develop an efficient computational algorithm that searches for the best peptide binder to a particular target. The biomarker chosen for initial study is Troponin I, a 210-amino acid protein that is elevated in the blood of heart attack victims and is a general indicator of muscle fatigue. The approach is based on a search algorithm already developed in the PIâ€™s group. An iterative procedure, which involves sequence mutation moves during which the peptideâ€™s backbone conformation is held fixed, and peptide backbone conformation moves during which the peptide sequence is held fixed, is used to arrive at the binding sequence and conformation that has the lowest energy. The search algorithm is supplemented by explicit-solvent atomistic simulations to determine the free energies of the top scoring peptides. The search algorithm was used to identify sixteen 12-residue peptides that bind to Troponin I. The starting sequence was a peptide, P0, discovered using phage display. Explicit-solvent simulations of the peptide-Troponin I complex were used to further screen the candidate peptides. A variety of experimental techniques (Surface Plasmon Resonance, Quartz Crystal Microbalance, Dot Blots) were used by our AFRL collaborators to assess the binding capabilities of the top three candidate peptides, P2, P3 and P5 as well as the original peptide, P0. The binding affinity of P2 was extremely strong at 1.53x 10^{-12}(M), compared to that of the original peptide, P0, at 6.15 x 10^{-9}(M).

543869 Reflections from Four Decades of Collegiality, Collaboration and Competition in AIChE

Abstract:

It is fitting to celebrate women in chemical engineering to mark the 20th anniversary of the Womenâ€™s Initiative Committee of the AIChE. I will mark this celebration by reflecting on my nearly forty years in the AIChE. Much has changed in these four decades and some things remain the same. As a group, chemical engineers are collegial, collaborative and competitive and these traits have shaped our discipline.

541939 Heterogeneous Catalyst Design at the Single Atom Limit for Efficient Chemicals Production

Abstract:

Supported single metal atoms are a new class of catalysts in which precious metals can be used at the ultimate efficiency limit. While great strides have been made in demonstrating the potential of single-atom catalysts for many industrial reactions, there remains much debate in the literature over the nature of the active sites. Atomically dispersed supported metal catalysts offer better resource utilization and new prospects for low-cost fuel processing and green chemicals production. Single metal atom catalysts stabilized on various supports will be discussed in this presentation, drawing examples from a variety of reactions, including the low-temperature CO oxidation and water-gas shift reactions, methanol steam reforming, and selective hydrogenation and dehydrogenation reactions on single-atom alloys.We will demonstrate how reaction mechanisms involving single metal atoms/cations, transcend support structure and composition, as long as the metal atom site is allowed to form stably. A unique â€œsignatureâ€ of the metal (Au, Pt, Pd, etc.) at the atomic state is preserved, distinct however from the corresponding extended metal surface.Selective dehydrogenation of alkanes, catalyzed by single-atom Pt/Cu support (single-atom alloy) will be presented. The activity is much higher than that of Cu, while the catalyst is coke-resistant under realistic conditions, unlike monometallic Pt.. Exploiting these promising new catalysts for the above and other reactions is a main goal of work carried out in our lab at Tufts.

542399 Chemical Process Development – Even for Reaction Engineers, it Ain't Just Kinetics

Abstract:

In his classic textbook, Scott Fogler characterizes reaction engineering with six fundamental pillars that support six overarching rooftops. Each pillar participates in the observed chemical reaction. In the rush to meet business imperatives, it is too easy to focus on the expected operating conditions and not consider how the pillars behave beyond that window. We will review a number of real (but sanitized) industrial processes where use of the appropriate tools and methodologies led to real process understanding that translated to improved process profitability.

542036 Celebrating Women in Chemical Engineering: Past and Present

Abstract:

In June of 2005, six women graduated from the California Institute of Technology and became the first and only all-female class of chemical engineers to graduate from a co-ed American university. Together, these women spent four years navigating the undergraduate chemical engineering curriculum at Caltech â€" growing and maturing as a team without male chemical engineering classmates. The Caltech 6 met in September of 2001 as an ethnically and geographically diverse group of women, and from there, they went on to defy all odds.

Engineering remains a male-dominated field; in classes, in companies and throughout the industry. In this world of men, there has only ever been one entirely female class of chemical engineers, and the Caltech 6 is it. Each woman has overcome her own individual struggles, from those that were self imposed to those born of societal presumptions. These challenges were easier to overcome because the Caltech 6 had the support system of women helping women. Their joint education was never a competition â€" it was only ever a collaboration. In this presentation, the undergraduate experiences of the Caltech 6 will be highlighted and the impact of this highly unusual chemical engineering undergraduate education on their post Caltech lives will be explored.

541749 Engineered Models of the Gut-Brain Axis

Abstract:

The gut-brain-axis is a complex bi-directional communication pathway between the gastrointestinal tract, the enteric nervous system (ENS), and the central nervous system (CNS) that is implicated in not only gastrointestinal function, but also cognitive tasks like memory and decision making. Gastrointestinal flora has also been implicated in alterations of brain function and behavior, however, mechanisms behind the gut-to-brain communication remain poorly understood. To investigate the mechanisms for epithelial/neural interactions in the gastrointestinal tract and understand the impact of alterations in neural activity in response to intestinal contents, we are developing in vitro humanized culture models of the enteric-gut axis. These platforms, termed â€œOrgans-on-chips,â€ have generated interest from academia and industry as these physiological models may augment drug and basic biological discoveries. However, the lack of rapid, scalable, and facile manufacturing techniques may limit the widespread use of organs-on-chips. Here I will discuss a novel laser-cut and assembly-based fabrication method for simple, and cost-effective thermoplastic organ-chips. It has also been proposed that seeding patient derived cells will enable personalized medicine, but current intestine-on-a-chip models utilize immortalized cells and do not include support cells such as enteric neurons. Finally, I will discuss the culture and differentiation of a primary, human epithelial monolayer from intestinal organoids for future on chip studies that recapitulate the heterogeneous gut population, and the impact of trophic cross-talk between the epithelium and enteric populations in static models.

541374 Designer Nanoplexes for Delivery to Targeted Tissues

Abstract:

The generation of LbL nanoparticles that can directly target a specific tissues such as metastatic tumors, the lungs, or other key organs is dependent on the nature of the outer LbL layer, and the biodistribution of the nanoparticle is highly dependent on net surface charge, degree of hydration, and type of polyelectrolyte bilayer pair that is adsorbed as the final layers on the nanoparticle. Recent work in which these nanoparticle systems are designed for optimized uptake by advanced serous ovarian cancer cells will be discussed, and the use of these approaches to deliver combination siRNA/chemotherapy or drug inhibitor combination therapies or to generate imaging systems and theranostic nanoparticles will be addressed. The potential to target other tissues using designed nanoplex systems is discussed, in particular for the case of targeting cartilage to address the early stages of post-traumatic osteoarthritis. The manipulation of outer surface charge and polymer chain functionality, as well as the ability to design these layered nanoscale complexes to respond to micro-environment cues to achieve controlled biodistribution and uptake to targeted cells in vivo will be described. Finally, the layer-by-layer approach can be used to generate finely tuned release surfaces that can release small molecule, proteins, nucleic acids and other biologic drugs over sustained time periods, and with significant control of release characteristics. This approach is particularly attractive for the delivery of proteins and nucleic acids such as siRNA. New polyplexes designed to complex mRNA with its capping protein, or siRNA with the Ago2 component of the RISC complex combine protein and RNA in stabilized complexes that greatly enhance translation, and exhibit promise for treatments including cancer vaccines.

541750 Biomedical Applications of Emulsion Templating

Abstract:

Emulsion templating is a recent fabrication method used to generate porous materials with exceptional control of pore architecture afforded by modulating emulsion variables. Our laboratory has developed a number of applications of polymerized high internal phase emulsions (polyHIPEs) in tissue engineering including injectable bone grafts, emulsion inks, and porous microspheres. First, fumarate-based polyHIPEs were developed that demonstrated a unique combination of properties for use as injectable, high porosity bone grafts. These materials can be stored up to one year without detriment to properties and then injected with rapid cure. Although polyHIPE tissue engineering scaffolds were first proposed in 2000, this is the first time a biodegradable polyHIPE was fabricated that cures at body temperature to a high porosity foam with compressive properties in the range of trabecular bone. Modulation of polymer chemistry has been used to generate a range of mechanical properties from rigid bone scaffolds to elastomeric, soft tissue grafts. Recently, we have taken advantage of the shear-thinning aspects of the HIPEs prior to cure to generate emulsion inks for 3D printing. These emulsion inks can be used to generate complex grafts with hierarchical porosity. Combination with thermoplastic filament printing permitted the fabrication biomimetic bone grafts with improved compressive properties without sacrificing permeability. Finally, a dual fluidics setup was used to fabricate porous microspheres for controlled release of bioactive factors. This solvent-free method provides advantages over current microsphere fabrication strategies including in-line loading of growth factors to improve loading efficiency. Overall, emulsion templating has demonstrated great potential to facilitate repair in a wide range of clinical specialties.

542396 AIChE's First Female Member – An Unsung Trailblazer of Chemical Engineering

Abstract:

In honor of WIC's 20th anniversary, a presentation of AIChE's first female member and her substantial impact. Margaret Hutchinson Rousseau was a trailblazer accomplishing many first including being the first email member of AICHE and she had a significant impact in process development & plant design in several industries including the first penicillin deep tank fermentation plant. The AIChE Institute Award for Lifetime Achievement by a Woman Chemical Engineer honors her memory.

541751 Silica Nanoparticles Act As Permeation Enhancers to Enable Oral Protein Delivery

Abstract:

Oral delivery of macromolecular drugs is a major unmet need in biomedicine. Poor absorption across the intestinal epithelium poses the greatest challenge to clinical translation. Here, we show that silica and other anionic nanoparticles facilitate the intestinal absorption of macromolecules, including insulin. Particles improved permeability across Caco-2 intestinal epithelial monolayers, with smaller (â‰¤ 100 nm) and more negative particles being more effective. In mice, 50 nm silica particles enhanced the absorption of 4,000 MW dextran and insulin in a dose-dependent manner. Furthermore, oral co-administration of silica nanoparticles and insulin-loaded capsules reduced blood glucose levels in healthy mice for more than 10 hours at insulin doses as low as 10 U/kg. Nanoparticle action is due to integrin binding, which activates myosin light chain kinase and rearranges the tight junction protein, ZO-1. Together, these data suggest that inexpensive silica nanoparticles enable true oral delivery of macromolecular therapeutics by reversibly opening intestinal tight junctions and without damaging tissue.

541748 The Role of Chemical Engineers in Pharmaceutical Development

Abstract:

Particle technology is a vast field that involves systems containing particles or powders. The pharmaceutical industry uses this technology when developing formulations such as aspirin, ibuprofen, and acetaminophen. Pharmaceutical companies are under tremendous pressure to make pharmaceutical production more efficient and more economical.

Granulation is often used in the design of specific pharmaceutical dosage forms as a means of achieving specific granule attributes. Pharmaceutical granulation dates back to the mid-1800s when the tablet press was invented. For some materials, granulation requires the usage of a binder or excipient that is capable of providing adhesive and cohesive forces for the particles. The binders are important in the granule formation and mechanical strength of the granule. Excipients vary in bonding tendencies and additive properties. Some excipients are known to cause adverse effects in select pharmaceutical formulations. More study is needed to characterize the binding performance of well used excipients and possible replacement materials.

AUTHOR INDEX